ASTRONOMY 120
LAB MANUAL

Revised Printing

Kelli Spangler

Montgomery County Community College

Kendall Hunt
publishing company

Kendall Hunt
publishing company

www.kendallhunt.com
Send all inquiries to:
4050 Westmark Drive
Dubuque, IA 52004-1840

Contents

*Adapted from Ferguson, Dale C. Introductory Astronomy Exercises, 2 edition
‡ Adapted from Fun in the Sun, Harvard University
◊ Adapted from University of Michigan Astronomy Department
⤢Adapted from RCottrell Meteor Crater Lab

Lab 1: The Celestial Sphere

Introduction:

The celestial sphere is a convenient way of representing the night sky as seen at any latitude and time of year. We will explore the properties of the celestial sphere and learn how they indicate what we observe in the night sky.

By learning how to use the celestial sphere, we will be able to do the following:

- Understand the concept of the *celestial sphere* and its defining components: *north and south celestial poles, celestial equator, zenith, horizon, ecliptic.*
- Be able to estimate *right ascension (RA)* and *declination (Dec)* of celestial objects from a celestial sphere.
- Be able to determine the constellations seen at particular latitudes.

Materials:

- Celestial Sphere

Procedure:

A. Declination and Right Ascension:

Declination is the sky's equivalent of latitude: position on the sky relative to the equator. Locate the lines of declination that run up and down, north and south, of the equator in units of 10 degrees.

Right Ascension is the sky's equivalent of longitude: position on the sky relative to 0 RA. These are in units of 15 degrees or 1 hour.

1. Locate the North and South Celestial Poles on the celestial sphere and record their declinations below.

    a. Dec NCP: _____
    b. Dec SCP:_____

2. Locate the Celestial equator on the celestial sphere and note its declination below.
    a. Dec Equator: _____

3. Locate the lines of RA on the celestial sphere. Find 0 RA. How many hours are represented on the entire globe? _____

B. *The Ecliptic:*

The ecliptic is the apparent path the sun takes through the night sky. This apparent motion is really the earth's orbital motion projected into the night sky.

1. Locate the Ecliptic on the celestial sphere and notice how it moves north and south of the celestial equator over the course of one full year.

2. Find the ecliptic on the 4 special days of the year (defined below) and record the Right Ascension (RA) and Declination (Dec) for the sun on those days.

   a. Vernal Equinox: The sun is on the equator moving north
   b. Summer Solstice: The sun is furthest north of the celestial equator
   c. Autumnal Equinox: The sun is on the equator moving south
   d. Winter Solstice: The sun is furthest south of the celestial equator

	Approx. Date	Right Ascension	Declination
Vernal Equinox			
Summer Solstice			
Autumnal Equinox			
Winter Solstice			

3. In what constellation is the sun located on the following days:

   a. July 5 _____

   b. January 20 _____

   c. November 10 _____

4. What zodiacal sign are you? _____

Let's double check: Your birthday_____
 Your sign from the old range:_____

Your sign from the new range:_____

Just recently, the signs have been updated to take precession into account and are more aligned with native cultures like the Mayans. Below are the new definitions of the zodiac signs: are you still the same sign you thought you were?

Sign	Old Range	New Range
Capricorn	12/21 – 1/20	1/20 – 2/15
Aquarius	1/21 – 2/20	2/16 – 3/10
Pisces	2/21 – 3/20	3/11 – 4/17
Aries	3/21 – 4/20	4/18 – 5/12
Taurus	4/21 – 5/20	5/13 – 6/20
Gemini	5/21 – 6/20	6/21 – 7/19
Cancer	6/21 – 7/20	7/20 – 8/9
Leo	7/21 – 8/20	8/10 – 9/5
Virgo	8/21 – 9/20	9/16 – 10/29
Libra	9/21 – 10/20	10/30 – 11/22
Scorpio	10/21 – 11/20	11/23 – 11/28
Ophiuchus		11/29 – 12/16
Sagittarius	11/21 – 12/20	12/17 – 1/19

Q: Why might your birth sign be incorrect?

5. Fill in the following table:

Proper Name	Name (Bayer system)	Right Ascension	Declination
		$18^{hr}\ 37^{m}$	$+39°$
		$7^{hr}\ 45^{m}$	$+28°$
	α Cygnus		
	β Taurus		
Altair			
Sirius			

Analysis Questions:

1. Are there any stars that cannot be seen from the equator? Why?

2. Are there any stars that can never be seen at the north pole and why?

Lab 2: Kepler's Third Law

Introduction:

Kepler recognized that the solar system obeyed three laws of planetary motion:

1. The planets moved in an ellipse with the sun at one focus
2. The planets covered equal areas in equal time resulting in non-uniform motion or a varying speed throughout the orbit
3. The size of the orbit (semi-major axis) cubed was directly proportional the period of the orbit squared

The one downfall of Kepler's laws was that he could offer no explanation of why his laws held. It was not until Isaac Newton formulized the fundamental rules of physics and all motion when he quantified the concept of a force and specifically the force of gravity. He recognized a direct relationship between the motion of an object and the forces it experiences. With regard to gravity, he realized that the force's strength depended upon two major properties, the masses of the objects involved as well as the distance between them. Once connected to Kepler, we can recognize that the distance between the two objects was defined as the semi-major axis of the planet's orbit. Kepler's third law can now be corrected to account for the force of gravity represented by the masses of the two objects involved:

$$M_1 + M_2 = \frac{a^3}{P^2}$$

This equation will hold so long as we use the appropriate units:

Mass (M) in Solar Masses (or the mass of the sun) written as M_{sun}
Semi-major axis (a) in AU
Period (P) in yrs

We will use the corrected version of Kepler's third law to determine the masses of objects in our solar system, of two binary stars, and of the Milky Way galaxy.

By learning Kepler's laws, we will be able to do the following:

- Define and understand terms such as semi-major axis and period
- To understand how orbital motion can yield estimates of mass
- To understand the limitations of Kepler's law
- Reveal the presence of an unknown material, Dark Matter, around all galaxies

Conversions:

1 pc = 206265 AU
1 AU = 1.5 x 10^{11} m = 1.5 x 10^8 km
1 yr = 3.15 x 10^7 s

Procedure:

A. *The Mass of Saturn*

In order to determine the mass of Saturn, we need to know some information about objects which are under the influence of Saturn's gravity, its moons. We will use its largest moon, Titan which sits 1.22 x 10^6 km and has an orbital period of 15.94 days.

1. Convert the distance from titan to Saturn from km into m

a = _____ m

2. Convert the distance from Titan to Saturn from m into AU

a = _____ AU

3. Convert the orbital period of Titan from days to years.

P = _____ yr

4. Apply Kepler's third law to determine the combined mass of Titan and Saturn in solar masses.

$$M_1 + M_2 = \frac{a^3}{P^2}$$

$M_{saturn} + M_{titan} = $ _____ M_{sun}

5. Due to the large difference in mass between Saturn and Titan, we can effectively ignore the mass of Titan when evaluating the answer to 4 and use this as the mass of Saturn in its entirety. Knowing that the mass of the sun (M_{sun}) is 1.99×10^{30} kg, find the mass of Saturn in kilograms.

$M_{saturn} = $ _____ kg

6. How does the mass of Saturn compare to that of the Earth if the Earth has a mass of 5.98×10^{24} kg.

7. From more accurate calculations, we find that Saturn is 95.147 times more massive than the earth. What might be some sources of error to cause your answer to 6 to be different than this?

B. *The Mass of the Milky Way Galaxy*

Newton's laws account for the mass which determines the motion of an object by measuring properties of the motion like orbital velocity, period, and radius. When we look at the motion of an object that lies within a symmetrical mass distribution, like in a galaxy, only the mass which lies inside its radius contributes to its orbital motion. Using Kepler's law we can get an estimate of the mass of the Milky Way galaxy by studying the motion of the stars which orbit around it.

If we look at the sun, we find that it orbits the center of the Milky Way galaxy at a radius or semi-major axis of 9.1 kpc with an orbital speed or velocity of 250 km/s

      1.   Convert the semi-major axis of the sun from kpc to pc.

      a = _____ pc

      2.   Convert the semi-major axis of the sun from pc to AU.

      a = _____ AU

      3.   Calculate the circumference of the sun's orbital motion by multiplying the semi-major axis by 2π

      Circumference = _____ AU

      4.   Convert the circumference from AU to km.

      Circumference = _____ km

5. Divide the sun's circumference in km by the sun's orbital velocity around the center of the Milky Way to determine its orbital period in seconds.

P = _____ sec

6. Convert the orbital period of the sun from s to yr.

P = _____ yr

7. Use the semi-major axis of the sun in AU and the period of the sun's orbit in yr to determine the combined mass of the Milky Way galaxy and the Sun.

$$M_1 + M_2 = \frac{a^3}{P^2}$$

$M_{Milky\ Way} + M_{Sun} =$ _____ M_{sun}

8. Knowing that the sun is infinitesimally small compared the Milky Way Galaxy, we can use the answer to number 7 as the mass of the Milky Way Galaxy which lies inside the orbit of the Sun. To how many billions of sun-like stars does this mass correspond?

C. *The Discovery of Dark matter*

Current observations of galaxies have allowed astronomers to measure the orbital properties of other stars which lie further out from the galactic center than the sun. We have found that their velocities don't change as we expected as we look further from the galactic center, with stellar speeds remaining relatively constant despite their increasing distance from the galactic center. According to Kepler and Newton, we should expect an object's velocity to decrease as this distance increases. We will repeat the above procedure for a star orbiting the center of the Milky Way at a radius outside the Sun to determine the mass of the Milky Way which is not limited by the sun and its orbital properties.

Our new star of interest lies further out than the sun at a distance of 20 kpc and orbits only slightly faster than the sun with an orbital speed of 300 km/s.

1. Convert the semi-major axis of the star from kpc to AU.

      a = _____ AU

2. Calculate the circumference of the star's orbital motion by multiplying the radius or semi-major axis by 2π

      Circumference = _____ AU

3. Convert the circumference from AU to km.

      Circumference = _____ km

4. Divide the star's circumference in km by the star's orbital velocity around the center of the Milky Way to determine its orbital period in seconds.

P = _____ sec

5. Convert the orbital period of the star from sec to yr.

P = _____ yr

6. Use the semi-major axis of the star in AU and the period of the star's orbit in yr to determine the combined mass of the Milky Way galaxy and the Star.

$$M_1 + M_2 = \frac{a^3}{P^2}$$

$M_{Milky\ Way} + M_{Star} =$ _____ M_{sun}

7. Knowing that the star is infinitesimally small compared the Milky Way Galaxy, we can use the answer to number 7 as the mass of the Milky Way Galaxy which lies inside the orbit of the star. To how many billions of sun-like stars does this mass correspond?

8. What percentage of the mass calculated at 20 kpc lies between our two radii of interest (9.1 kpc and 20 kpc)?

9. We notice from the answers to the last two parts of this lab that the amount of mass greatly increases as we move outward in the galaxy from the center yet when we look at a picture, we fail to see this mass, noticing that the majority of the light of the galaxy comes from the center rather than the edges. What conclusion must we draw about the outskirts of the galaxy?

Pinwheel Galaxy: Chandra.si.edu

Lab 3: Spectrum Scavenger Hunt

Introduction:

Atoms are composed of two main parts: the nucleus where the neutron and proton reside and the electron cloud where the electrons reside. The electrons orbit the nucleus in orbital levels that correspond to a specific amount of energy. When we excite an electron, we give it enough energy to change its orbital or energy level to a higher one. When the electron returns to its natural or original energy level, it must return that energy in the form of a photon of light. The photon of light will have the energy value equal to the difference of the higher minus the lower energy levels of the electron. This energy will correspond to a unique and specific color represented by the wavelength of the photon.

The colors emitted by an atom can be seen when we take a spectrum, a special "picture" of an object taken with an instrument called a spectrograph. The spectrograph will break up the total light, a mixture of all colors, into its individual colors allowing astronomers to identify the composition, temperature, and a variety of properties based upon the energies associated with the wavelengths present in the spectrum.

There are three types of spectra: continuous (all colors are seen like a rainbow), emission (specific colors are seen as individual lines of color), absorption (looks like a continuous spectrum with specific colors removed which looks like dark bands).

The learning outcomes of this exercise is to:

- Understand how properties of atoms give rise to the wavelength of light
- Understand what a spectrograph is and how it works
- Identify and analyze atomic spectra
- Identify the composition of the light fixtures through the ATC and surrounding courtyard based upon the spectral lines the light source emits.

Equipment:

- Handheld spectrographs
- Light sources around campus as well as provided by professor

Procedure:

1. Use the hand-held spectrographs to identify the colors/wavelengths present in the light sources listed below. Once you have identified the wavelengths present, use the table at the end of the lab to identify the composition of the light source.

Data:

Light source 1: classroom lights

Wavelength (nm)	Wavelength (nm)

Type of spectrum (circle): emission   absorption   continuous

Light source 2: Light outside classroom door

Wavelength (nm)	Wavelength (nm)

Type of spectrum (circle): emission   absorption   continuous

Light source 3: Light Bulb provided by professor

Wavelength (nm)	Wavelength (nm)

Type of spectrum (circle): emission   absorption   continuous

Light source 4: Spectral Tube 1 _____

Wavelength (nm)	Wavelength (nm)

Type of spectrum (circle): emission   absorption   continuous

Light source 5: Spectral Tube 2 _____

Wavelength (nm)	Wavelength (nm)

Type of spectrum (circle): emission   absorption   continuous

Light source 6: Spectral Tube 3 _____

Wavelength (nm)	Wavelength (nm)

Type of spectrum (circle): emission   absorption   continuous

Light source 7: Spectral Tube 4 _____

Wavelength (nm)	Wavelength (nm)

Type of spectrum (circle): emission   absorption   continuous

Light source 8: the sun

Wavelength (nm)	Wavelength (nm)

Type of spectrum (circle): emission   absorption   continuous

Light source 9: Running lights by TV/Comp area in ATC (Night only)

Wavelength (nm)	Wavelength (nm)

Type of spectrum (circle):   emission   absorption   continuous

Light source 8: lights outside ATC in courtyard (Night only)

Wavelength (nm)	Wavelength (nm)

Type of spectrum (circle):   emission   absorption   continuous

Analysis Questions:

1. What is the main difference between the spectrum of the classroom lights and the spectrum of the "old fashioned" light bulb provided by professor? Why do you think there is such a difference?

2. What is the main difference between the spectrum of the classroom lights and that of the spectral tubes? Why do you think these are different?

3. What is different about the spectrum of the sun? What kind of spectrum is it?

4. What does the spectrum of the running lights look light? What can we conclude about how that bulb works?

5. What kinds of spectra are the lights outside the ATC? Was there a difference between the spectra of the parking lot lights vs. the old fashioned light bulb? (night section only)

Table 1: Common spectral lines			
Wavelength (nm)	Element	Wavelength (nm)	Element
350	Hydrogen	560	Argon/ Mercury/Neon
365	Chromium	565	Oxygen/ Carbon
380	Hydrogen	570	Neon
385	Helium	577	Mercury
387	Carbon	580	Neon
395	Oxygen	587	Helium/ Carbon
400	Helium	590.0	Sodium/ Mercury/Neon
405	Mercury	595	Argon
407	Carbon	600	Carbon/Neon
415	Nitrogen	605	Neon
420	Hydrogen/Argon	610	Neon
430.8	Iron	611	Argon
434	Hydrogen	615	Oxygen/Neon
436	Mercury	620	Neon
438	Helium	625	Argon/ Mercury/Oxygen
440	Hydrogen/Argon/Oxygen	640	Nitrogen/Argon
447	Helium	650	Argon
450	Mercury	656.3	Hydrogen
455	Helium	657	Carbon
460	Mercury	660	Argon/Mercury/Oxygen
468	Helium	665	Neon/Oxygen
471	Helium	667	Helium
475	Neon	670	Hydrogen/Argon/Neon
477	Carbon	678	Carbon
490	Hydrogen/Neon/Oxygen	680	Argon
491	Mercury	685	Neon
501	Helium	700	Helium
505	Mercury/ Carbon	705	Neon
510	Helium/Neon	710	Argon/ Carbon
517.3	Magnesium, Iron	715	Neon
520	Nitrogen	720	Mercury
525	Oxygen/Neon		
527	Iron		
538	Carbon		
540	Oxygen		
546	Mercury		
550	Argon/Oxygen		

Lab 4: Craters and the Moon

Introduction:

This lab will focus on the origin and properties of craters. In the first part of the lab, we will focus on the types of craters different meteors can make. We will be able to explore how different sizes of meteors as well as different types of geology can impact the properties associated with craters. In the second part of the lab, we will apply this information to the moon to discover what information its craters can provide. This analysis will help to formulate the theory of the origin of the moon.

The learning objectives of this lab are:

- To identify the basic mechanism of making a crater
- To connect the crater left behind to the impactor object
- To identify the different terrains on the moon
- To connect the ideas of cratering to the craters observed on the moon
- To develop a theory for the development of the moon since its formation

Part A: Basic Cratering

Materials and Equipment

- Plastic bins
- rulers
- stations of: birdseed, flour, sand
- 3 different sized spherical objects

Procedure: There will be 3 "stations" at which you will perform a crater making experiment.

1. Using a ruler, measure the diameter of each object. The diameter is the distance across the middle of the sphere, from one side to the other. Divide this number by 2 to get the radius of each object. Record this in the data tables provided.

2. Using the scale, measure the mass of each object. Record this in the data tables provided

3. Determine and record the density (ρ) of each object by using the mass (m) and radius (r) in the following equation: $\rho = \dfrac{m}{\frac{4}{3}\pi r^3}$

4. Go to station 1. Shift the box from side to side to evenly distribute the material.

5. Drop one of your "meteors" into the box by holding the object *from the same height* for each trial. Simply let the object fall: DON'T THROW OR ADD ADDITIONAL FORCE to the meteor. Allow gravity to do all the work for you.

6. After the "meteor" impacts the material, carefully remove the object without disturbing the "crater" left behind.

7. Measure the diameter and depth of the first "crater" by measuring the distance across the center of the depression and the distance to the bottom of the depression. Record them in the data table. Be very careful not to disturb the material while making your measurement.

8. Drop the object two more times, carefully removing the object each time. You should have three "craters" to measure for each object. Measure the diameter and deprth of the subsequent two craters, writing each measurement in the data table.

9. Calculate the average crater diameter and depth by adding up the three measurements and then dividing your answer by three. Write the answer in your data table.

10. Prepare your box for the next "meteor" by shaking it from side to side to even out the material until it is smooth and level.

11. Repeat steps 4–10 for all of your objects, each time recording the diameter of the three "craters" and the average in the data table.

12. Repeat steps 4-11 for the other 2 stations.

Object Information			
Object	**Radius (m)**	**Mass (kg)**	**Density (kg/m^3)**
S			
B			
G			

Station 1:									
Object	**Diameter (m)**			**Average Diameter (m)**	**Average radius (m)** $r = \dfrac{AvgDiam}{2}$	**Depth of Crater (m)**			**Average Depth (m)**	**Average volume (m^3)** $V = \pi r^2 * depth$
S										
B										
G										

Station 2:							
Object	Diameter (m)			Average Diameter (m)	Average radius (m) $r = \dfrac{AvgDiam}{2}$	Depth of Crater (m)			Average Depth (m)	Average volume (m³) $V = \pi r^2 * depth$
S										
B										
G										

Station 3:							
Object	Diameter (m)			Average Diameter (m)	Average radius (m) $r = \dfrac{AvgDiam}{2}$	Depth of Crater (m)			Average Depth (m)	Average volume (m³) $V = \pi r^2 * depth$
S										
B										
G										

Analysis Questions:

1. What correlation can you make between crater volume and the properties of the meteor objects? In other words, does the mass, radius, or density of the impacting object seem to be the most important property when it comes to making craters?

2. How did the surface material affect the craters the objects made? Can you describe how these observations can be used to determine geological properties of impacted objects?

Part B: The Moon

Materials and Equipment

- Atlas of the Moon (Antonin Rukl 2004)

Procedure/Analysis:

I. Lunar surfaces

1. Near Side: Take a look at the full moon on page 190, there seems to be two distinct types of lunar surfaces. What are they? Try to use simple words to describe them.

2. Mare refers to the dark material that we easily see with the naked eye. Prior to telescopic images, we thought these were bodies of water, hence the name. Is there any evidence that these are not bodies of water?

3. If the surface of the moon was cratered at a constant rate, what can we determine about the relative ages of the two types of surfaces?

4. Far Side: On page 191 we find a map of the far side of the moon. How is it different than the near side?

II. Craters

5. Take a look at the pictures on Pg 196. What seems to be a common feature in the center of craters?

6. Look at Tycho on page 197 (pic 11). What do you see emanating from it? What do you think this might be? Did you observe something like this in part A?

7. Look at Mare Crisium on page 207 (pic 50). What common lunar feature does this mare resemble? Is there anything different about it compared to the other lunar features?

8. Look at Mare Imbrium on page 190. Name the features surrounding it?

9. Does Mare Imbrium seem similar to Mare Crisium? Why?

10. Some of the maria have narrow depressions running across them called rilles as seen in pictures 22-26 on page 200-201. What Earth-like features do these resemble?

11. What do you think caused the rilles on the moon?

12. From lunar samples, we believe that the lunar lava was thin and free flowing. What conclusion can we draw regarding the origin of these Maria?

13. Does the origin of Maria correspond to the theory discussed in class regarding the origin of the Moon? Why?

Lab 5: Measure the Size of the Sun

Introduction:

When you look at the sun in the sky, it seems rather small considering it's a star. However, we realize that this is only because of its distance from the earth. How large it appears in the sky must be connected to both, the real size (R_{sun}) and its distance from the earth ($a = 1.5 \times 10^{11}$m). We will determine the actual size of the sun by taking measurements based upon how large it appears in the sky.

The Learning objectives of this lab are:

- To understand how distance makes objects appear smaller
- How to measure the apparent size of an object

Materials:

- 8.5 x 11 sheet of paper
- Meter stick
- Tape
- Index card
- Pen/pencil

Procedure and Analysis:

NOTE: DON'T LOOK DIRECTLY AT THE SUN

1. On the sheet of paper, draw two parallel lines VERY CLOSE TOGETHER (approximately 1 or 2 mm). Measure this separation (d) in mm and place it in the line below.

    d = _____2_____ mm

2. Poke a tiny hole in the index card with the pin **(this must be a tiny hole or your data will be very inaccurate).**

3. Fold the index card over on one end

4. Tape bent portion to the top of the meter stick at the 0m mark (figure 1).

5. Hold the paper with the parallel lines up against the edge of the meter stick as shown in figure 1.

6. Make sure the light created by the pin-hole falls on the side of your paper with the parallel lines.

31

7. Move the paper up and down the meter stick until the sunlight fills the space between the lines on the paper.

8. Record the position on the meter stick (L) where the paper was aligned in step 7 above in mm.

L = _____ mm

Figure 1

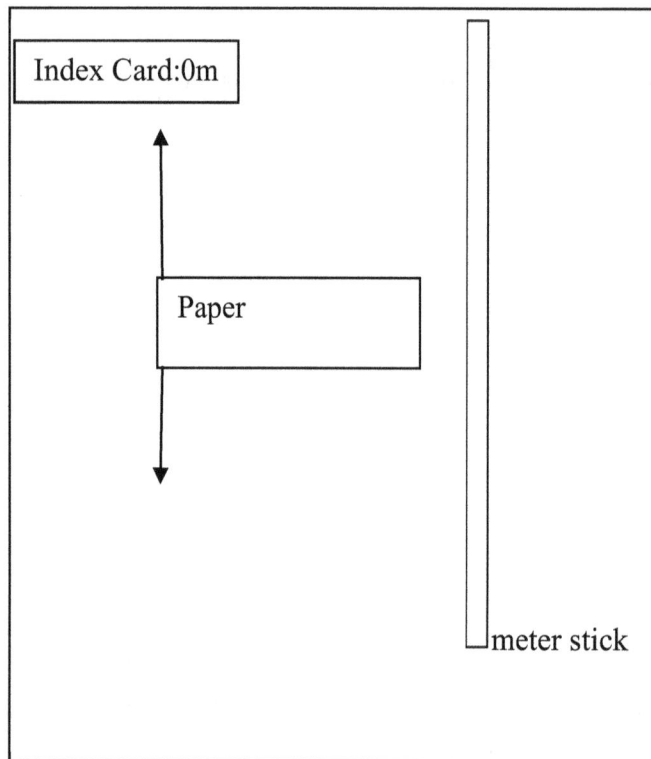

9. Because the size of the sun (R_{sun}) is directly related to the apparent size and distance to the sun, we can set up a simple relationship between the measurements we took. First, calculate the ratio below. *Note:* the units cancel once you do this division leaving this number without units. Please don't convert mm to m.

L/d = _____

10. The ratio of L/d will also equal the ratio a/R_{sun} where we already know the semi-major axis of the earth (a). Carry down your answer from part 9 and copy it in here.

$a/R_{sun} = L/d$

$a/R_{sun} =$ _____

11. Rearrange the above relationship to solve for R_{sun}. Remember to substitute for the semi-major axis of the earth (a).

$R_{theoretical} =$ _____

12. What is the percent error in your answer? Use $R_{sun} = 1.2 \times 10^9$ m?

$$\frac{R_{sun} - R_{theoretical}}{R_{sun}} \quad *100 \ =$$

13. What might be your sources of error for this lab?

Lab 6: Determining Parallax

Introduction:

The idea behind parallax is quite simple: it is the apparent change in an objects position due to the motion of the earth. This jump in position is directly related to how far the object is from the earth. This process is used by the human brain to determine distance to objects and is limited by the small separation between the eyes. By increasing this separation to the orbit of the earth and replacing the eye with a space based telescope, our measurements of parallax greatly improved. In this lab, we will determine the distance to an object across the room by using triangulation:

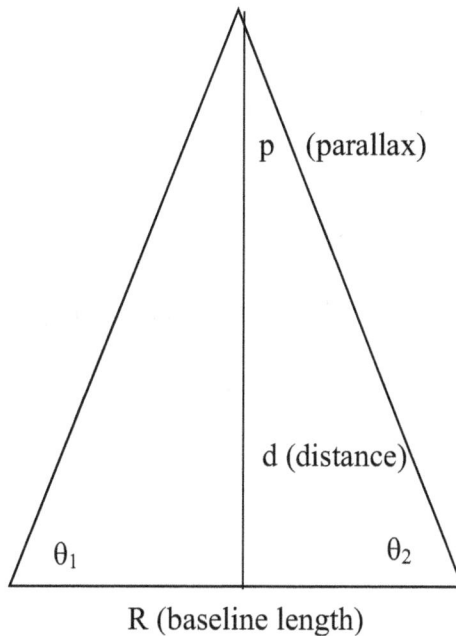

Because the line of sight generates a triangle (triangulation), it will allow us to use basic trig functions to determine the distance to the object via the following equation:

$$\text{Tan}(p) = \frac{R/2}{d}$$

The learning objectives of this exercise are:

- To understand how parallax works
- To connect the idea of parallax with distance

Equipment:

- Table surface
- Paper
- Meter stick
- Protractor

Procedure:

1. Write down the distance from the object to you (about 1-2 meters works well).

   Actual d_{actual} = _____ cm

2. Draw a base line on your paper that is 25 cm in length and is parallel to the long side of the paper. Mark the 0 cm and 25 cm positions at the ends of the line.

   Baseline R = _____**25**_____ cm

3. Position the paper on your table so that the line you drew is **perpendicular** (90 degrees) to the object in the front of the room.

4. Lay the meter stick on the table so that one end is on the 0 cm mark of your baseline. Then, line up the meter stick so that the object is seen at the end of your line of sight along the meter stick.

5. Draw a line tracing the meter stick/line of sight on the paper (sight line 1).

6. Repeat this sighting at the 25 cm mark and draw a second line (sight line 2).

7. Measure the two angles of your triangle using a protractor.

   θ_1 = _____ degrees

   θ_2 = _____ degrees

8. Calculate the third angle by adding up the other two and subtracting them from 180 degrees.
   $\theta_1 + \theta_2$ = _____

   $$p = \frac{180 - (\theta_1 + \theta_2)}{2}$$

   p = _____ degree

9. Calculate the distance to the object by using the parallax and the baseline via the following equation (be sure your calculator is set to degrees):

$$d = \frac{R/2}{\text{Tan}(p)}$$

$$d_{\text{theoretical}} = \text{_____} \text{ cm}$$

Analysis Questions:

1. How did your theoretical answer in step 9 compare to the actual distance from step 1?

$$\frac{(d_{\text{actual}} - d_{\text{theoretical}})}{d_{\text{actual}}} * 100$$

2. What do you think might have been your sources of error?

Lab 7: The Hertzsprung-Russell Diagram

Introduction:

The Hertzsprung-Russell Diagram (HR) is a very useful diagram that we will use to organize and describe stars throughout their evolution. It is a graph of the luminosity of stars versus their temperature. From the position of a star on the HR diagram, we can determine the radius (size), mass, color, temperature, distance, brightness, and evolutionary stage of a star.

By using the HR diagram, we will be able to:

- Determine physical properties of stars such as their radius, temperature, mass, and luminosity
- Determine the evolutionary stage of a star
- How to tell a constellation apart from a cluster of stars

Procedure and Analysis:

I. Basic HR properties:

Using the figure provided by the instructor, answer the following questions:

1. Where are the hottest (bluest) stars found? The coolest (reddest)?

   Bluest:

   Reddest:

2. Where are the largest stars found? The smallest?

   Largest:

   Smallest:

3. Where are the most massive stars found? The least massive?

   Most massive:

   Least massive:

4. Where are the brightest stars found? The dimmest?

Brightest:

Dimmest:

5. Label these properties on the blank HR diagram in figure 1

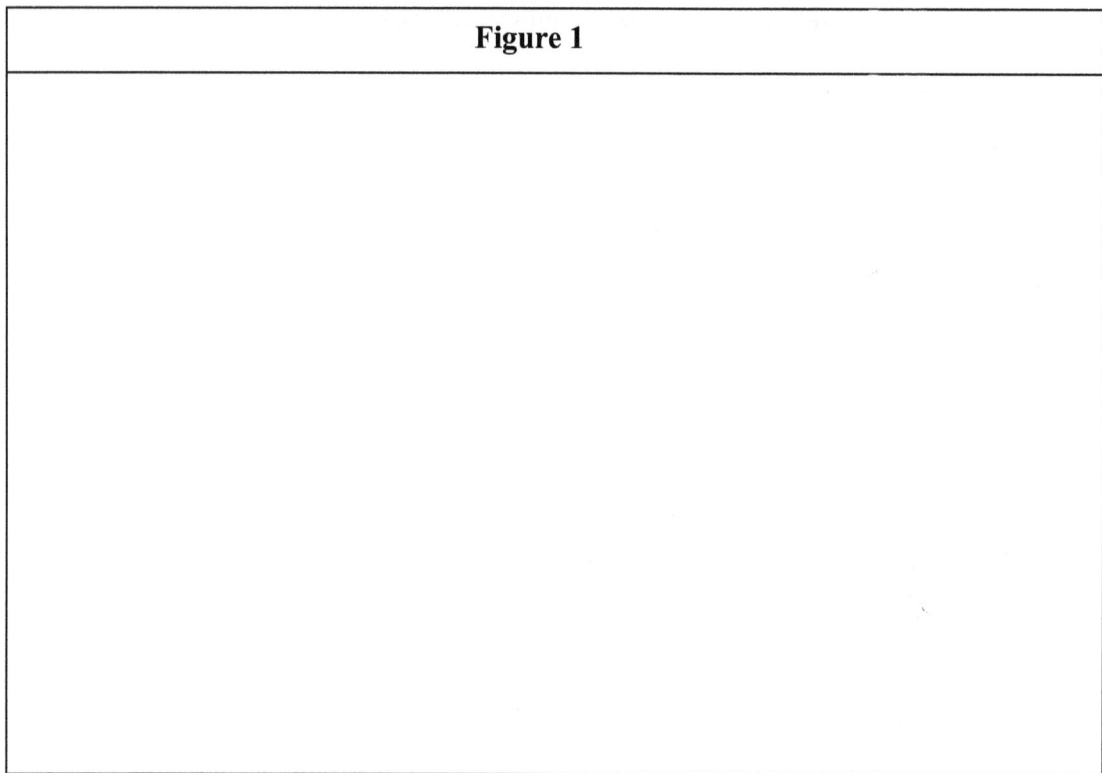

Figure 1

II. Stars of Taurus

1. Plot the stars of Taurus (not including the pleiades) provided in Table 1 on Figure 4. Please notice that the negative numbers are at the top of the Y-Axis since negative magnitudes denote brighter stars.

2. What star is the hottest? Coolest?

3. What star is the brightest? Dimmest?

4. What star is the most massive? Least massive?

5. What star is the largest? Smallest?

6. What star is the bluest? Reddest?

7. Where do all of the stars of Taurus fall on your plot?

8. What does this tell you about these stars?

9. Look at the plot of the cluster housing the Pleiades provided by the instructor (Figure 5). It was created using *apparent magnitude vs. color* not using absolute magnitude vs. color. Using the apparent magnitude information of the Pleiades from table 1, circle those stars we commonly refer to as the "seven sisters" on figure 5.

10. Where are the seven sisters found in the plot? What does this mean about them?

11. What is the overall shape of this plot? What name would we give to this shape?

12. What is the major difference between the plot you made (figure 4) and figure 5?

13. What is the major difference between the apparent magnitude and absolute magnitude?

14. Because the plot for the Pleiades (figure 5) was made with apparent magnitude, it must be dependent upon the same property you discovered in question 13. When we look at this plot, we find that it makes a straight line, indicating that all the stars in the Pleiades have a direct relationship with respect to this specific property. What is the only direct relationship they can have?

15. Because of this direct relationship, what can we conclude about the Pleiades?

16. What do you think would happen if we re-plotted the stars of Taurus by apparent instead of absolute? Would we expect to see a straight line?

Figure 4

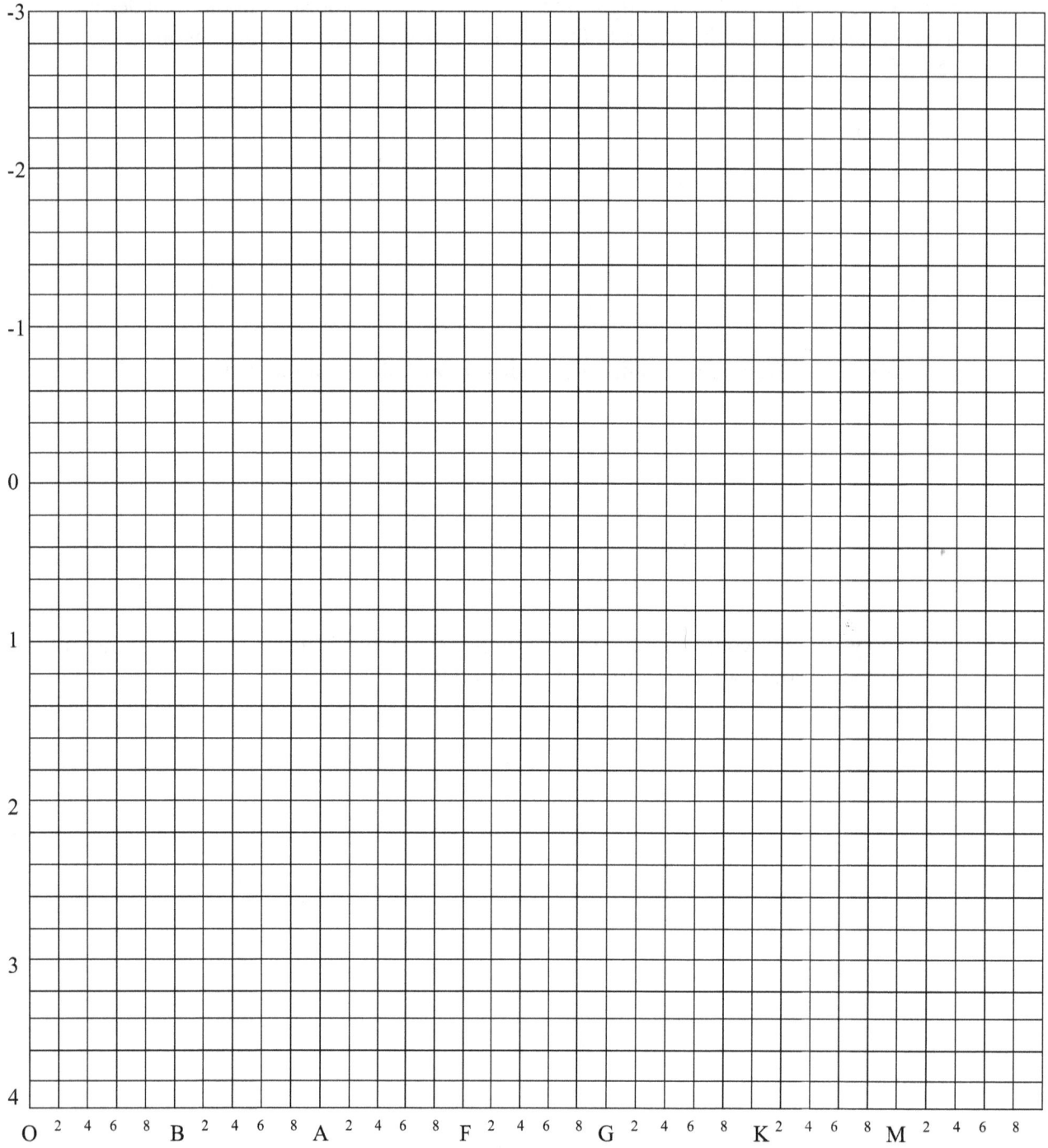

Table 1

Name	Apparent Magnitude	Absolute Magnitude	Color
α	0.87	-0.63	K5
β	1.65	-1.37	B7
η	2.85	-2.41	B7
ζ	2.97	-2.56	B4
θ^2	3.4	0.10	A7
λ	3.41	-1.87	B3
ε	3.53	0.15	K0
o	3.61	0.45	G8
γ	3.65	0.28	G8
ξ	3.73	-0.44	B9
δ^1	3.77	0.41	G8
θ^1	3.84	0.42	G7
ν	3.91	0.92	A1
f	4.14	-1.08	K0
κ^1	4.21	0.85	A7
μ	4.27	-1.35	B3
τ	4.27	-1.18	B3
Pleiades			
16	5.45		B7
17	3.69		B6
19	4.29		B6
20	3.86		B7
21	5.75		B8
23	4.16		B6
27	3.62		B8

Lab 8: Classification of Galaxies

Introduction:

In order to study galaxies, we must first categorize them by some method. Hubble first introduced a classification scheme based upon visual properties of galaxies. He developed three primary types of galaxies: Spiral (S), Elliptical (E), and Irregular (Irr). The goal of this lab is to determine the visual properties of galaxies that can be used to classify them. Then we will apply this information to a cluster of galaxies to help determine the origin and evolution of galactic type.

The learning outcomes of this lab are:

- To identify a spiral, elliptical, and irregular galaxy by using visual properties only
- To develop a theory of elliptical galaxy creation by looking at galactic environments.

Materials:

- Kit pictures (9 in total)
- The Hubble Atlas of Galaxies

Procedure:

Visual Properties:

1. Take a look at kit picture 2. This is an example of a spiral galaxy. What are its visual properties?

2. Which other kit picture(s) are examples of spiral galaxies? Do you see any differences between them?

3. Take a look at kit picture 1. This is an example of an elliptical galaxy. What are its visual properties?

4. Which other kit picture(s) are elliptical?

5. What is different about kit picture 5? Does it appear to have the same visual properties as either a spiral or elliptical galaxy? Of what type of galaxy must you conclude this is an example?

6. In the Hubble Atlas, turn to page 2 and take a look at M87 (small image in top right corner). What is different about its visual properties?

7. While on page 2, turn your focus to galaxies NGC 750/751. How does the atlas explain what we see here? What does that mean?

8. Turn to page 38 in the atlas and take a look at the LMC. As what kind of galaxy would you classify it?

9. If I told you that this was a dwarf galaxy in a spiraling orbit about the milky way, how might it help to explain its appearance?

Environment:

Look at the picture of the Virgo Cluster provided by the instructor (Figure 1). It is an irregular cluster located approximately 14.9 Mpc from our local group. There are 7 clear galaxies in the picture as labeled on the key (Figure 2).

1. Classify the types of galaxies seen above by circling E (elliptical), S (spiral), or Irr (irregular). Also refer to the images located at the end of the lab for closer inspection of each galaxy.

    1. E S Irr
    2. E S Irr
    3. E S Irr
    4. E S Irr
    5. E S Irr
    6. E S Irr
    7. E S Irr

2. Relative to the cluster, where do you find the elliptical galaxies?

3. Relative to the cluster, where do you find the spiral galaxies?

4. What is strange about galaxy 5? To what can you attribute this property?

5. According to the most recent theories regarding galactic evolution, elliptical galaxies are believed to be the result of galactic collisions/mergers while spiral galaxies are considered to have been isolated from any major catastrophes since their birth. Does the distribution of elliptical versus spiral galaxies support this hypothesis?

6. How do you think galaxy 5 fits into this new theory?

Galaxy 1: M86 by Hubble Space Telescope

Galaxy 2: M84 by NOAO

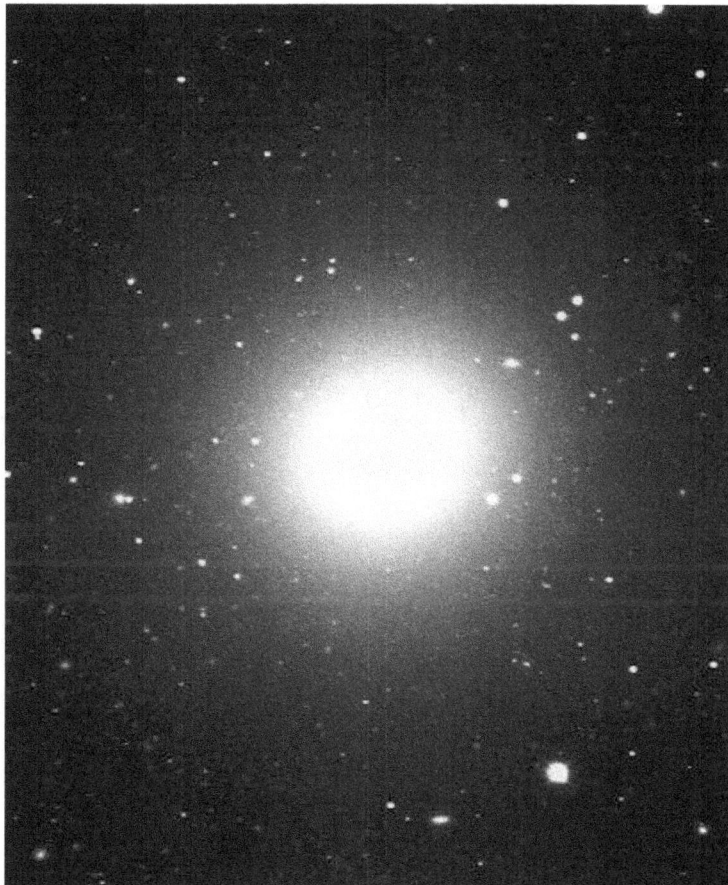

Galaxy 3: NGC 4388 by Hogg/SDSS

Galaxy 4: NGC 4425 by Digitized Sky Survey

Galaxy 7: NGC 4402 by Emil Ivanov Astrophotography

Movies

We will use movies toward the end of the semester as a broad way to introduce or delve deeper into topics we discuss. In order to guide you through the movie, allowing you concentrate on the important, relevant information, you will fill out the accompanying assignment as you watch. We will then sit as a large group, go over your answers, and further discuss the importance of the information you just garnered.

Always remember that movies not only deliver information from a new perspective but they also provide graphical images to help you envision some arcane topics such as dark matter, dark energy, feeding or non-feeding super-massive black holes, parallel universes, and the space-time continuum just to name a few.

All of these movies can be found through the library's "Films on Demand" digital streaming service.

        To access them:

1. Log into your portal
2. Click the "Libraries" link
3. Click on the "AV" tab
4. Click on The "Films on Demand" link found directly beneath the search bar to be taken to their login page
   a. You may need to use your school information to log into the site
5. It is usually easiest to search by the title but you can also search using the submenus found at the bottom of the page.

Lab 9: The Planets

The beginning of the movie goes very quickly through the first 8 questions. Please try to follow along as best you can.

1. How long ago did the solar system form?

2. What two forces are at work in the solar nebula?

3. Which force won? What happened..what shape came out of this result

4. What object formed first? Where?

5. The inner planets have what characteristics in common?

6. What is the asteroid belt?

7. The outer giant planets have what characteristics in common?

8. What are comets mostly made of?

Mercury:

9. Mercury most resembles what other solar system body?

Venus:

10. What is Venus' atmosphere like and why is the planet so hot?

11. What is Venus lacking that Earth has in abundance?

Earth:

12. How did the moon form?

Mars:

13. Where can we find most of the water on Mars?

14. How many moons does Mars have?

Jupiter:

15. What is Jupiter mostly made of?

16. What is the great red spot on Jupiter?

17. How many large moons does Jupiter have? What are some of their properties?

Saturn:

18. What is the most prominent feature of Saturn? What is it made of?

19. What is the largest moon of Saturn like?

Uranus:

20. What is the difference between Uranus' composition and that of Jupiter/Saturn?

21. What is the most unusual thing about Uranus?

Neptune:

22. Of what is Neptune made?

23. What does the great dark spot resemble? What happened to it?

1. How does the surface of the sun behave?

2. Of what elements is the sun made?

3. In what large objects are stars born?

4. What is produced as the cloud collapses?

5. How does the solar system's birth fit into this model?

6. What is happening in the central core of the sun?

7. Why is it necessary for the core of the sun to be so hot and dense?

8. What element is produced from this H fusion?

9. What is a sun spot?

10. How does the equator of the sun rotate differently than the poles?

11. What is the cycle of sun spots?

12. What is the connection between sun spots and flares/coronals mass ejections?

13. What force is responsible for all this behavior?

14. What does the magnetic field store or contain?

15. How do the loops of magnetic field get affected by the differential rotation of the sun?

16. What happens when a loop snaps?

17. What happens when the material released by the sun hits the earth's magnetic field?

18. In about 5 billion years, what will happen to the sun?

19. What will the body of the sun do as it reacts to the loss of H?

20. What will we call the sun then?

21. What will happen to the core of the sun?

Lab 11: Who's Afraid of the Big Black Hole?

1. Why is it hard to find black holes?

2. Where do we believe black holes come from?

3. What did Doug Leonard observe that made him think he saw the birth of a black hole?

4. How did Einstein describe the behavior of gravity (General Relativity GR)?

5. Einstein says mass affects space-time how?

6. What was predicted by GR that was troubling?

7. Ultimately, what would happen if you fell into a black hole?

8. What is the point of no return for a BH?

9. How does this explain how light is "sucked" into a BH?

10. Inner Horizon: What ultimately happens as you approach the center?

11. At the heart of the BH, why do Einstein's math/theories break down?

12. What is the singularity?

13. What did astronomers find that proved BH could exist?

14. Ultimately how do astronomers detect BH's including in the center of the Milky Way?

15. What mass did Prof. Genzel calculate for the Milky Way's SMBH?

16. What is the major problem with GR and the natural world? What theory deals with this troubled area?

17. Why is it so important to use Quatum Mechanics (QM) simultaneously with GR for BH's to work?

18. What is so different about the quantum world compared to the world you experience and why is it so readily accepted?

19. What is one thing QM can't describe? What do we call the possible theory that can?

20. What happened when physicists tried to combine QM and GR?

21. What is the one other place where the laws of nature fail? Why do the laws again fail?

Lab 12: Most of the Universe Is Missing

1. How was dark matter first observed?

2. Where do we find all this dark matter?

3. What are some of the explanations given regarding what dark matter is?

4. What is one major argument for why dark matter doesn't necessarily exist?

5. What observations led to the idea of dark energy?

6. How do these combine to make the current model of cosmology?

7. What is the CMB (or WMAP) and how did its observation support the current model of cosmology?

Lab 13: Elegant Universe: Einstein's Dream

1. What two things did Newton unite?

2. What was the major flaw with Newton's theory of gravity that Einstein addresses?

3. How did Einstein change the model of the universe?

4. How did Einstein explain the behavior of gravity?

5. What two things did Maxwell unite?

6. Why did Einstein admire Maxwell?

7. What was one example that illustrated how weak gravity was compared to the other forces of nature?

8. Why did Einstein reject the ideas of quantum mechanics?

9. What are the 4 basic forces of nature?

10. What parts of nature does quantum mechanics govern?

11. What parts of nature does general relativity govern?

Lab 14: Elegant Universe: String's the Thing

1. What is a string?

2. How is it different than a particle?

3. How do physicists use particles to explain force?

4. What is the problem with gravity under this explanation?

5. How small are strings?

6. How does the size of the string "help" string theory?

7. What does the vibration of the string determine?

8. What does string theory do to QM's erratic behavior?

9. How can string theory unite the two branches of physics?

10. What is the major difficulty in proving string theory?

11. Does string theory only need the 4 dimensions of space we already have or more than that?

12. What is a clear example that more than 4 dimension are possible?

13. What are the 2 varieties of extra dimension?

14. Why can't we see these extra dimensions?

15. What do the 6 extra dimensions look like?

16. How does the shape of the dimensions affect the string?

17. How do the 20 fundamental constants of nature depend on strings?

18. Why do some physicists reject the idea of string theory?

19. How many versions of string theory once existed?

20. What were the two major types of strings mentioned?

Lab 15: Elegant universe: The 11th Dimension

1. What was the new name given to the theory Ed Whitten revealed by merging the 5 versions of string theory?

2. If string theory is correct, how many dimensions are there in the universe?

3. Where does the 11th dimension exist? What is it called?

4. What do we mean by parallel universes?

5. Why can't we interact with a parallel universe?

6. What is one example of how weak gravity in the universe is?

7. How does the movie explain particles of matter and quantum mechanics vs particles of gravity? There are two analogies made. Please explain one.

8. How could we possibly interact with a parallel universe?

9. How could parallel universe interaction explain the big bang?

10. Why is the new particle smasher (large hadron collider) and Fermi lab important? What are they looking for and how could it support string theory?

Astronomy 120
Outdoor Lab Observing Sheets

Date: _____

Object Name:_____

NAKED EYE, BINOCULAR, AND TELESCOPIC SKETCHES ARE REQUIRED

Analysis questions:

1. What color(s) does it seem to be?

2. What kind of an object is it? E.g. is it a star, planet, moon etc.

3. The Moon: Are there any features on the surface such as craters or mountains? Name and sketch these features.

4. Planets and sun: Do you see any atmospheric structure? Sketch any structure that you see in one of the provided sketch diagrams.

Sketches:

You are to do one sketch for each instrument and/or magnification. These must include: naked eye, binocular, and telescopic sketches where the telescope can offer multiple magnifications (up to 2 on the deck plus observatory).

Each sketch is to be done in the "field of view" circle provided where you are to fill the circle with what you observe in the eyepiece.

Make sure to label the magnification for telescopic sketches with the focal length of the eyepiece.

Naked Eye

Binocular

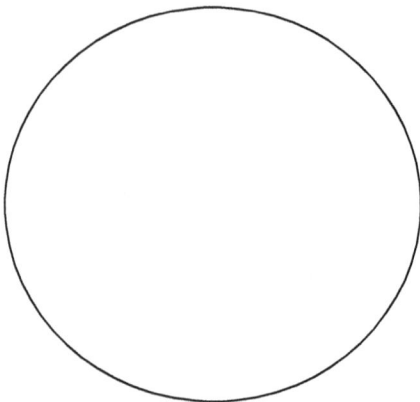

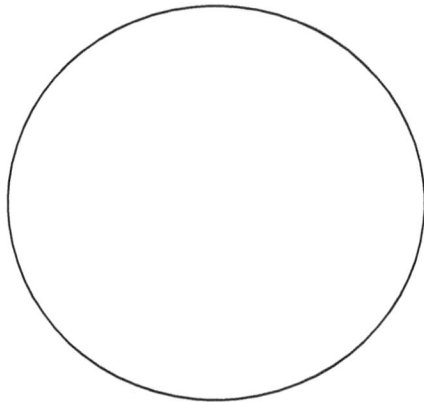

Telescopic 1

Eyepiece focal length: _____

Telescopic 2

Eyepiece focal length: _____

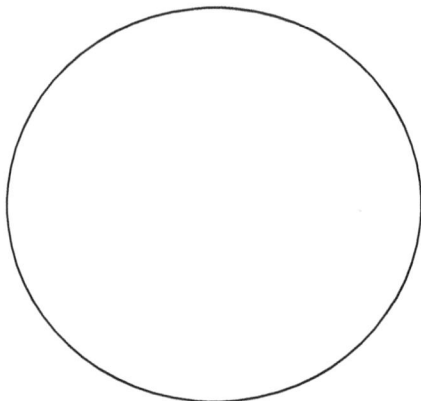

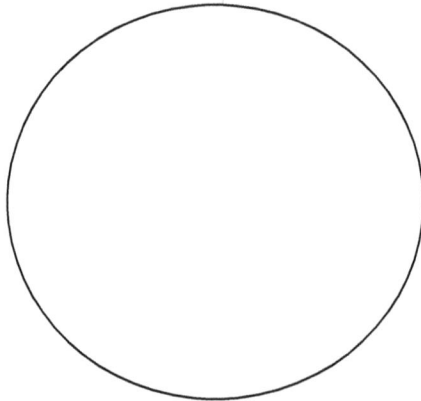

5. Planet: Does the object have moons? If so, sketch all that you see.

6. In what constellation is the object located? Is it rising or setting?

7. In what direction is the objected located: N, S, E, W, SE, SW, NE, NW.

Telescopic 3

Eyepiece focal
length:

_____

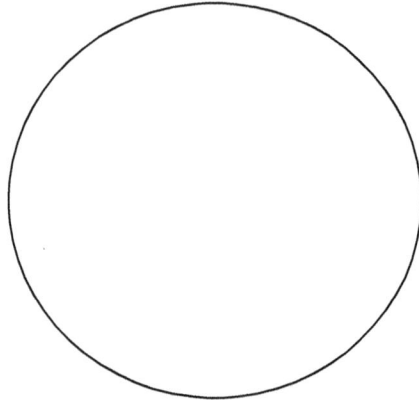

Astronomy 120
Outdoor Lab Observing Sheets

Date: _____

Object Name:_____

NAKED EYE, BINOCULAR, AND TELESCOPIC SKETCHES ARE REQUIRED

Analysis questions:

1. What color(s) does it seem to be?

2. What kind of an object is it? E.g. is it a star, planet, moon etc.

3. The Moon: Are there any features on the surface such as craters or mountains? Name and sketch these features.

4. Planets and sun: Do you see any atmospheric structure? Sketch any structure that you see in one of the provided sketch diagrams.

Sketches:

You are to do one sketch for each instrument and/or magnification. These must include: naked eye, binocular, and telescopic sketches where the telescope can offer multiple magnifications (up to 2 on the deck plus observatory).

Each sketch is to be done in the "field of view" circle provided where you are to fill the circle with what you observe in the eyepiece.

Make sure to label the magnification for telescopic sketches with the focal length of the eyepiece.

Naked Eye	Binocular

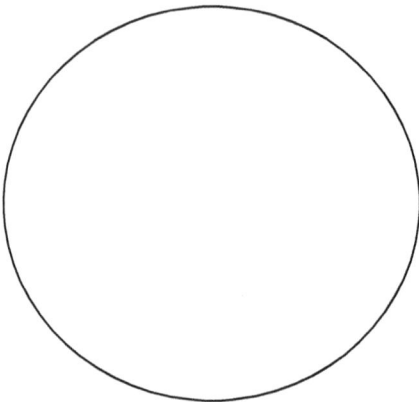
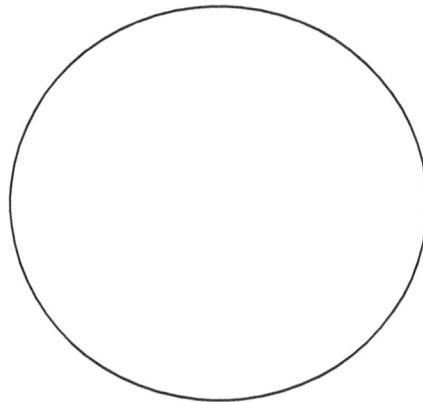

Telescopic 1 Eyepiece focal length: _____	Telescopic 2 Eyepiece focal length: _____

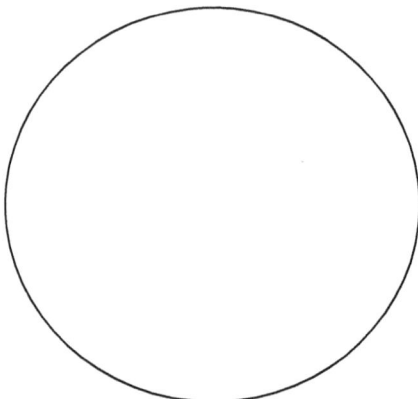
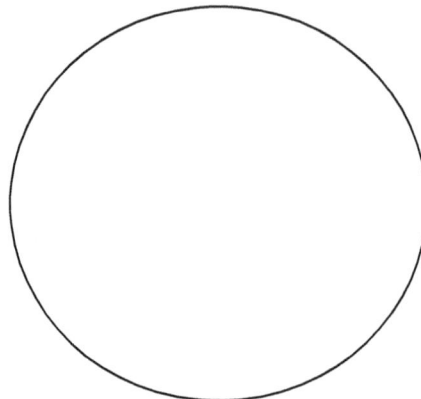

5. Planet: Does the object have moons? If so, sketch all that you see.

6. In what constellation is the object located? Is it rising or setting?

7. In what direction is the objected located: N, S, E, W, SE, SW, NE, NW.

Telescopic 3

Eyepiece focal
length:

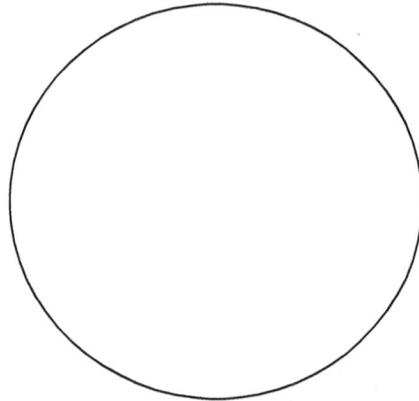

Date: _____

Object Name:_____

NAKED EYE, BINOCULAR, AND TELESCOPIC SKETCHES ARE REQUIRED

Analysis questions:

1. What color(s) does it seem to be?

2. What kind of an object is it? E.g. is it a star, planet, moon etc.

3. The Moon: Are there any features on the surface such as craters or mountains? Name and sketch these features.

4. Planets and sun: Do you see any atmospheric structure? Sketch any structure that you see in one of the provided sketch diagrams.

Sketches:

You are to do one sketch for each instrument and/or magnification. These must include: naked eye, binocular, and telescopic sketches where the telescope can offer multiple magnifications (up to 2 on the deck plus observatory).

Each sketch is to be done in the "field of view" circle provided where you are to fill the circle with what you observe in the eyepiece.

Make sure to label the magnification for telescopic sketches with the focal length of the eyepiece.

Naked Eye		Binocular

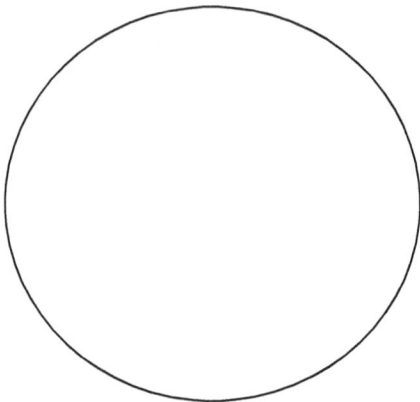
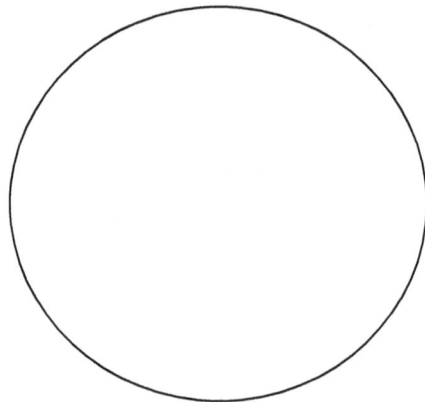

Telescopic 1		Telescopic 2
Eyepiece focal length: _____		Eyepiece focal length: _____

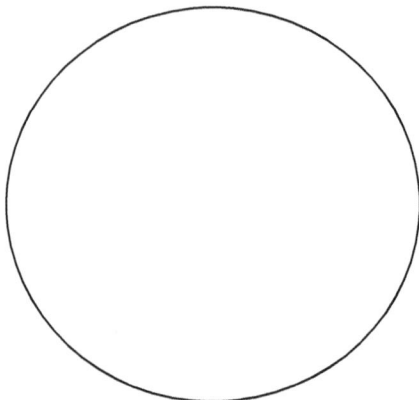
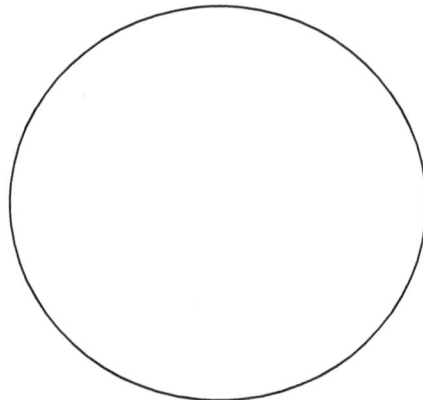

88

5. Planet: Does the object have moons? If so, sketch all that you see.

6. In what constellation is the object located? Is it rising or setting?

7. In what direction is the objected located: N, S, E, W, SE, SW, NE, NW.

Telescopic 3

Eyepiece focal
length:

Astronomy 120
Outdoor Lab Observing Sheets

Date: _____

Object Name:_____

NAKED EYE, BINOCULAR, AND TELESCOPIC SKETCHES ARE REQUIRED

Analysis questions:

1. What color(s) does it seem to be?

2. What kind of an object is it? E.g. is it a star, planet, moon etc.

3. The Moon: Are there any features on the surface such as craters or mountains? Name and sketch these features.

4. Planets and sun: Do you see any atmospheric structure? Sketch any structure that you see in one of the provided sketch diagrams.

Sketches:

You are to do one sketch for each instrument and/or magnification. These must include: naked eye, binocular, and telescopic sketches where the telescope can offer multiple magnifications (up to 2 on the deck plus observatory).

Each sketch is to be done in the "field of view" circle provided where you are to fill the circle with what you observe in the eyepiece.

Make sure to label the magnification for telescopic sketches with the focal length of the eyepiece.

Naked Eye		Binocular

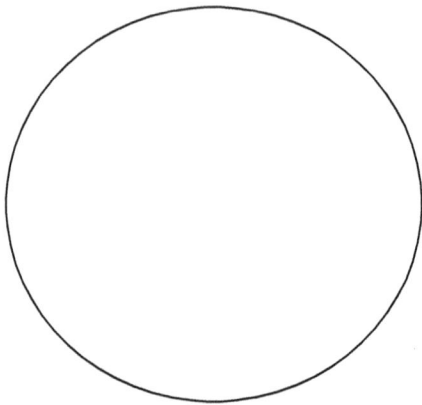 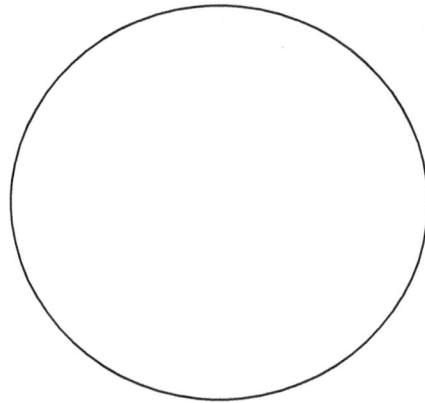

Telescopic 1		Telescopic 2
Eyepiece focal length: _____		Eyepiece focal length: _____

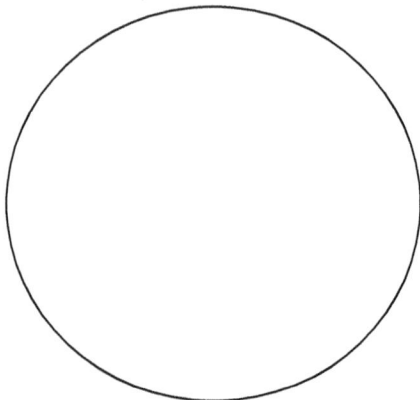 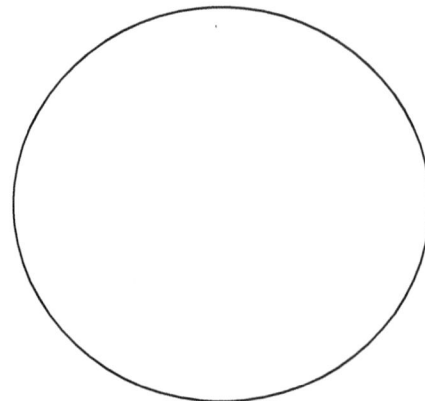

5. Planet: Does the object have moons? If so, sketch all that you see.

6. In what constellation is the object located? Is it rising or setting?

7. In what direction is the objected located: N, S, E, W, SE, SW, NE, NW.

Telescopic 3

Eyepiece focal
length:

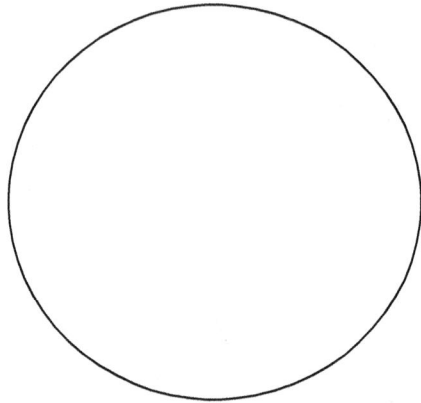

Date: _____

Object Name:_____

NAKED EYE, BINOCULAR, AND TELESCOPIC SKETCHES ARE REQUIRED

Analysis questions:

1. What color(s) does it seem to be?

2. What kind of an object is it? E.g. is it a star, planet, moon etc.

3. The Moon: Are there any features on the surface such as craters or mountains? Name and sketch these features.

4. Planets and sun: Do you see any atmospheric structure? Sketch any structure that you see in one of the provided sketch diagrams.

Sketches:

You are to do one sketch for each instrument and/or magnification. These must include: naked eye, binocular, and telescopic sketches where the telescope can offer multiple magnifications (up to 2 on the deck plus observatory).

Each sketch is to be done in the "field of view" circle provided where you are to fill the circle with what you observe in the eyepiece.

Make sure to label the magnification for telescopic sketches with the focal length of the eyepiece.

Naked Eye	Binocular

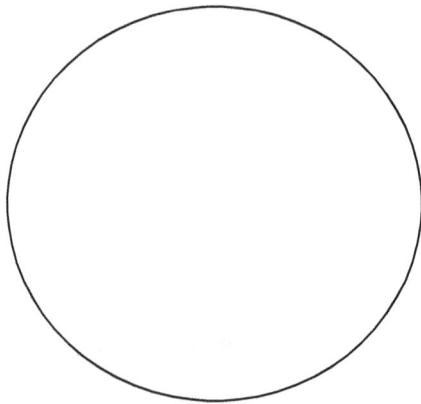
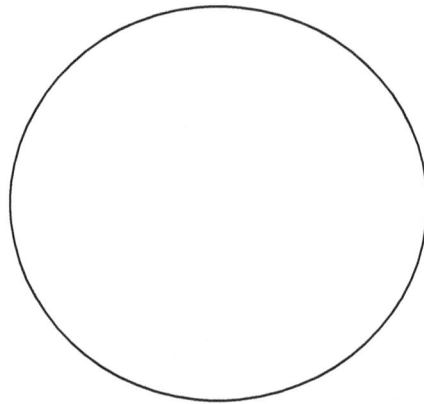

Telescopic 1	Telescopic 2
Eyepiece focal length: _____	Eyepiece focal length: _____

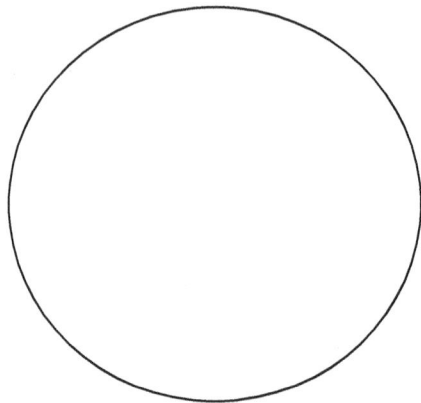
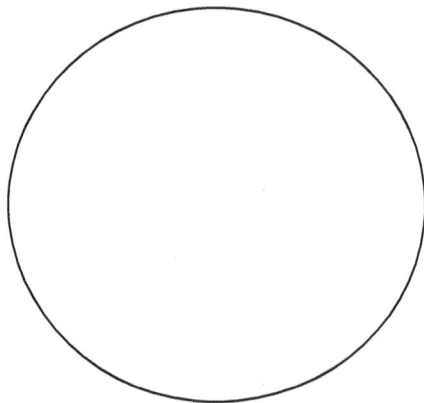

96

5. Planet: Does the object have moons? If so, sketch all that you see.

6. In what constellation is the object located? Is it rising or setting?

7. In what direction is the objected located: N, S, E, W, SE, SW, NE, NW.

Telescopic 3

Eyepiece focal
length:

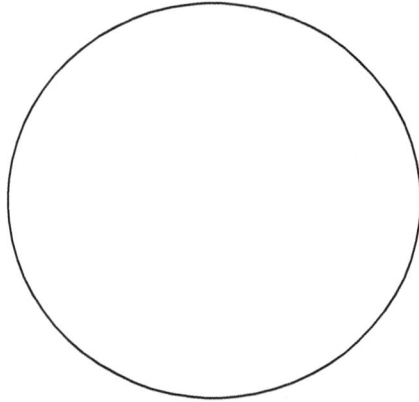

www.ingramcontent.com/pod-product-compliance
Lightning Source LLC
Chambersburg PA
CBHW082107210326
41599CB00033B/6617